Ian Carroll is a best selling author, with all of h
paperback and also on Kindle.

Ian is the author of the 'A-Z of Bloody Horror'
titles – *'Warning: Water May Contain Mermaids'*, *'Antique Shop'*, *'Clown in Aisle 3'* and *'Pensioner'*. Also the author of the horror books *'My Name is Ishmael'*, *'Demon Pirates Vs Vikings – Blackhorn's Revenge'*, *'The Lover's Guide to Internet Dating'* and *'Valentines Day'*.

He is also the author of the music books –
'Lemmy: Memories of a Rock 'N' Roll Legend' – which was a #1 in the UK, USA, Canada, France and Germany – *'Ronnie James Dio: Man on the Silver Mountain – Memories of a Rock 'N' Roll Icon'*, *'Leonard Cohen: Just One More Hallelujah'*, *'Music, Mud and Mayhem: The Official History of the Reading Festival'* and *'From Donington to Download: The History of Rock at Donington Park'*.

The First six Volumes of the *'Fans Have Their Say...'* series are also available which are:

KISS, AC/DC, BLACK SABBATH, GUNS N' ROSES, METALLICA & QUEEN

Ian also writes the history section for the Official Reading Festival music site in the UK and has attended the festival 31 times since 1983.

Ian lives with his wife Raine, two sons – Nathan & Josh and a jet-black witches cat called Rex - in Plymouth, Devon, UK.

www.iancarrollauthor.com
Facebook.com/iancarrollauthor (Various Book Pages as well)
ian@iancarrollauthor.com

End of an Era: Is This the End? Of Course Not, It's Just the Beginning...

© Ian Carroll 2019

ISBN-10: 9781093815849

No part of this publication can be reproduced in any form or by any means, electronic or mechanical – including photocopy, recording or via any other retrieval system, without written permission from the Author/Publishers.

All Photographs provided for this Project have been credited to to Ian Carroll and/or Karl Parsons.
Any Newspaper or TV stills remain copyright of the respective outlets.
All other Photographs remain the copyright of the various associated production and distribution companies and are presented here for educative and review purposes only and should not be reproduced in any way.

The End of an Era

Is This the End?
Of Course Not, It's Just the Begining

© Ian Carroll 2019

End of an Era: Is This the End? Of Course Not, It's Just the Beginning...

End of an Era: Is This the End? Of Course Not, It's Just the Beginning…

Foreword

We fought hard and long and sometimes into the night, but the Reel Cinema refused to move and the venue was closed for the final time on 28th February 2019 – The Gala Bingo Hall, which is also attached to the venue closed in August 2018.

The final performances in the evening of 'Alita Battle Angel' and 'Instant Family' were both very well attended and people stayed around afterwards to take photos and make the most of the final film showings at the Reel, for the time at least.

Our group – The Royal Cinema Trust – have met once a month for the last two years, gradually growing, but always built around the core of Karl, Tim, Hilary, Caroline, Nicki, Jessica and myself. We have fought to save the Cinema building, so although the Cinema has closed, the building is still there and at the moment there seems to be no plans to demolish it.

Ideally we want to see the building saved and used as a multi-use venue, with concert venue, theatrical stage, music practice rooms and a cinema – showing classic films, double bills and late night/all night film festivals.

So we are still campaigning to save the building, nothing has changed, we are still looking at getting funding through various options that have already been looked into and we have already drawn up detailed plans that will be presented to Plymouth City Council, finger crossed we will win in the end…

This Book, shows you behind the scenes at the venue and many places that you will have never seen, so have fun and hope you enjoy the Book.

50% of royalties from the sales will go towards our ongoing campaign.

Ian Carroll April 2019

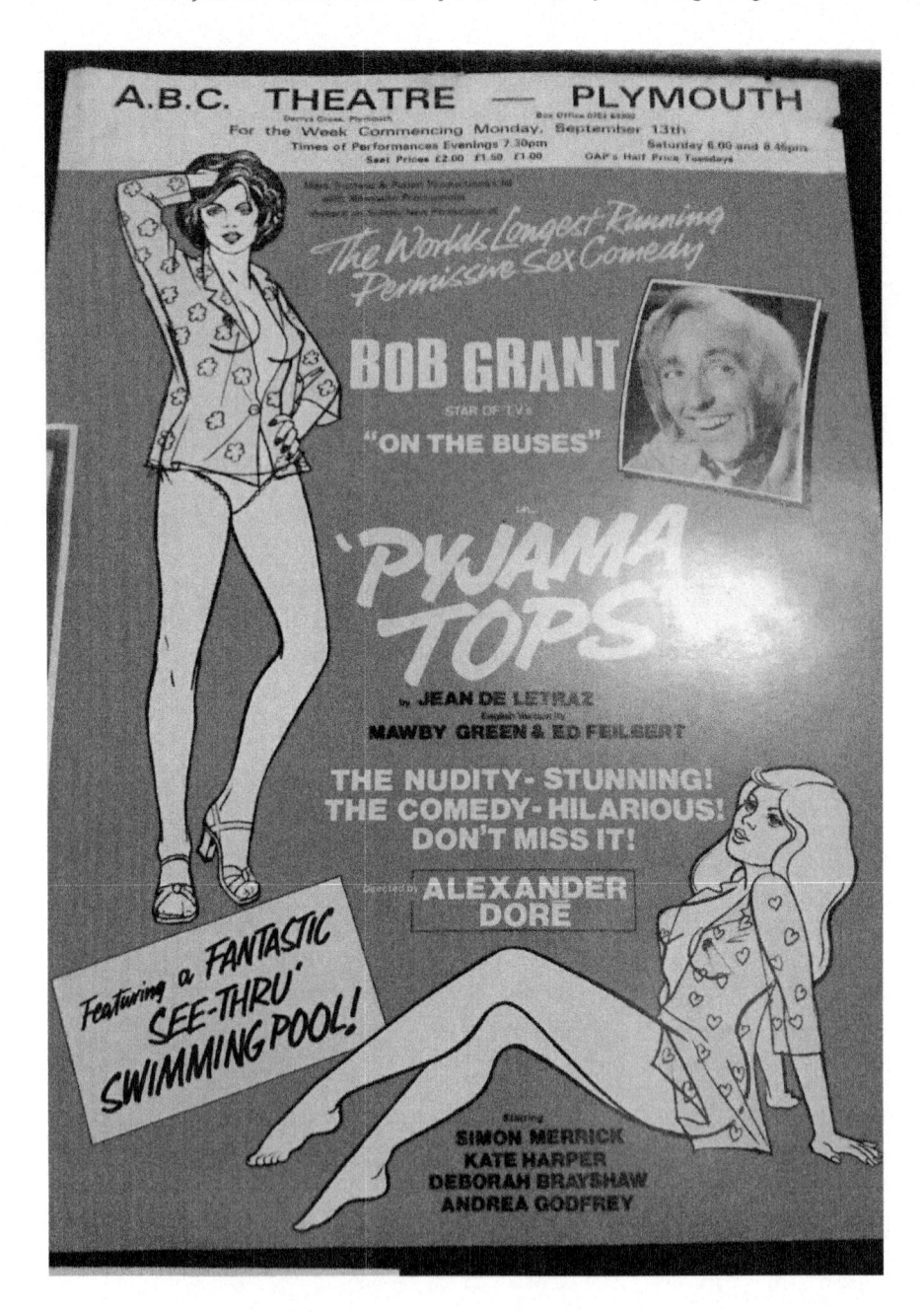

End of an Era: Is This the End? Of Course Not, It's Just the Beginning...

End of an Era: Is This the End? Of Course Not, It's Just the Beginning...

Introduction

(Previously Used in the Book – "The Last Picture House")

The Cinema
"The Royal Cinema opened on July 15th 1938, replacing the Theatre Royal which had stood on the site since 1813. It was designed by William Riddell Glen for Associated British Cinemas (ABC). The ABC chain had been founded in 1928 as part of British International Pictures by John Maxwell, and by 1937 was the second largest cinema chain in the world. Unlike it's competitors, this growth was accompanied by profits and annual dividends with no financial crises.

With 2,404 seats the Royal was unusually large for a provincial ABC, but the nearby Gaumont Palace at 2,300 seats and Regent at 3,250 seats also held large audiences, something which may reflect the sheer number of military personnel stationed in Plymouth in the 1930s. An Odeon planned for Union Street but never built would have had a capacity of 2,500.

Played at the opening by Wilfred Southworth, the Compton organ, regarded as one of the best ever fitted to a cinema became well known at the hands of Dudley Savage. He began at the end of 1938 at just 18 years old and remained the resident organist until 1976. Weekly BBC radiobroadcasts and album releases brought Dudley Savage and the 'Plymouth Sound' to a wide audience.

The Architect
William Riddell Glen, born in 1884 began designing cinemas in 1919 and became the in-house architect for the ABC chain in 1929. His contribution has been overlooked to some extent, with little mention of him in architectural literature compared with George Coles and Harry Weedon. However, it is Glen's skill in interior planning which has helped to keep the Royal open for so long, facilitating the subdivision of the building, which has helped it survive social changes.

The layout of all Glen cinemas was superb, with patrons moving forward from the entrance and pay boxes to their seats, rather than having to negotiate the passageways and winding staircases typical of 1930s cinemas. The spaces he created made his cinemas the most profitable part of the ABC circuit, and although all his interiors displayed the same skill, each was an individual design developed for its location. Glen's 1938 cinemas are regarded as among his best, with the Royal described by Eyles as 'Glen on top form'.

The War
Though all cinemas closed with the declaration of war on September 3rd 1939, they soon re-opened, and cinemas played an important role in entertaining and informing the nation. On March 20th 1941, the area around the cinema suffered massive damage with fire gutting the Royal Hotel and Assembly Rooms but the Royal only closed for a short period for repairs, probably thanks to its modern fireproof construction. Twyford,

in his description of the Plymouth blitz, said that

'Those modern buildings like the Royal, Odeon and Gaumont, and the 'ever faithful' Palace Theatre, received only slight damage and were able to continue their full programs within a short time of the heavy raids in 1941'

By 1943 a third of the population went to the cinema at least once a week, even with a 9.30pm curfew in place to save fuel and rubber. The war years were a golden era for British cinema, both in popularity and creativity.

There were 1,635 million cinema admissions in the UK in the first year of peacetime, with ABC having 18% of that total. About a third of 7-10 year olds were going to Saturday morning film clubs that year, including the 'ABC Minors' - which had been revived in October 1945. At the time film clubs were criticised by educationalists as bad for children's development- showing content which was too violent, showing too many imports, and being generally rowdy, charges which seem to be leveled at the entertainment of every generation of children. Indeed, 'juvenile delinquency' was a term which had been coined during the war due to limited parental supervision.

The Postwar Challenge

Pearl and Dean first appeared on screens in February 1952 with advertising being introduced to offset the revenue lost to falling attendances and reduced ice-cream sales after the abolition of sweet rationing. Attendances continued to fall and by the end of the 1950s TV had taken over as mass entertainment, with cinemas began closing down in significant numbers, a trend which continued over the next 20 years. Live shows were introduced at many of the larger ABC's, as a way to bolster income and between 1954 and 1976 a wide variety of music and entertainment acts appeared at the ABC, including 2 appearances by the Beatles.

During the 1970s while the total number of cinemas fell in the UK the number of screens rose sharply as a result of twinning and tripling conversions. With the Forum Devonport (1938) having closed in 1960 and the State in St Budeaux (1939) shutting in 1973, The ABC and Drake were converted to multi-screen cinemas at a similar time in the mid 1970's, securing their future at least in the short term. After a final live performance by Morecambe and Wise in 1976, The Stalls at the ABC were converted into a Bingo hall, at first known as the EMI Social club and now Gala Bingo. The Circle became screens 1 & 2 and Screen 3 was inserted in the double height foyer. Nationally attendances continued to fall and by 1980 cinema attendance was at 2% of the 1950 figure. However the Drake Odeon and the ABC were both getting Saturday children's matinee attendances of 800 in 1979- 3% of the UK total and showing that the city didn't necessarily mirror national trends, something which may explain the survival of both the ABC and Drake throughout the 1980's.

The Theatre District

Following the severe damage inflicted during the war the reconstruction of the city, first proposed in the 1943 Plan for Plymouth, has been described as 'nationally exceptional and significant' by Professor Jeremy Gould. The Royal Cinema was one of the very few pre-war buildings specifically included in the Plan for Plymouth, forming the nucleus of a proposed theatre district.

It was originally intended to build a new hotel on the site of Foulston's Hotel and Assembly Rooms, and WR Glen had drawn up detailed plans for this. The hotel would have adjoined the cinema, which explains the 'unfinished' corner on the left of the building where the bingo hall entrance now sits. However, following the blitz the vacant site left by the destruction of the original hotel was filled with Nissen huts which formed a temporary NAAFI until the Hoe Centre on Notte Street could be built.

By the 1980's the theatre district proposed by Abercrombie was complete, with the new Theatre Royal, The ABC Cinema, the Athenaeum, Westward Television/TSW and the Drake Cinema all in place. The Royal Cinema, Derry's Cross and the Bank pub are the only reminders of the pre-war street layout, with the site *of the old Royal Hotel now taken up by the Theatre Royal Car park and Princess Way. Sadly the theatre district has lost both Westward Television and the Drake Cinema, but the Royal Cinema, the Theatre Royal and the Athenaeum still entertain and educate.*

What's In a Name

*The cinema opened as the '**Royal**' in 1938, reflecting the history of the site. Neon lighting was fitted to ABC's from 1949, along with a redesigned ABC logo. Due to the loss of so many venues in the blitz the cinema served as a theatre as well as showing films, with the name '**Theatre Royal**' added to the canopy. This was short lived though and by 1958, in line with other ABC cinemas, the name 'Royal' was dropped and new corporate '**ABC**' branding applied, along with a new canopy which is still above the entrance today. Warner Bothers had controlled the ABC chain since 1946, but sold their stake to EMI in 1969 and the '**EMI**' logo was added to the existing ABC signage.*

*From the 1980s onwards the British cinema industry went through a complex series of changes in ownership and branding. ABC became '**Cannon**' in 1987 when Thorn EMI Screen Entertainment sold up to businessman Alan Bond, who then sold the chain on a week later for a multimillion pound profit. Cannon was taken over by Pathe in the early 1990s with* the '**MGM**' brand appearing on cinemas. Via a brief period under the ownership of Virgin the remains of the Odeon and ABC chains were merged and by 2000 the Odeon brand was applied to the remaining ABC's. However, a sole survivor, the Westover Bournemouth, remained an ABC until it's closure in January 2017. In Plymouth a management buyout had prevented the ingominy of Odeon branding, with the '**ABC**' name returning to the Royal in the mid 1990s. The current operator, Reel Cinemas, took over in 2006 and the cinema has been branded a '**Reel**' since then.

The Last Picture House

"Seven cinemas in Plymouth and three in Devonport were lost in the blitz and after the war the only new cinema to be built was the Drake- intended as the first in a chain of Twentieth Century Fox cinemas. This plan foundered and soon the Drake became part of the Odeon circuit, resulting in 3 Odeon's in close proximity. The Regent (an Odeon since 1940) soon closed down and was demolished, then following the closure of the other Odeon (formerly the Gaumont Palace) in 1980, Plymouth city centre was left with the ABC and the Drake. Further afield the Plaza on Bretonside (built in 1934) had closed in 1981 after its twilight years as an adult cinema, and the Belgrave in Mutley shut its doors in 1983.

With the opening of the Vue multiplex to the east of the city centre, the Drake sadly closed in 1999 and was demolished, though 'the Ship' survives, marooned on the casino building which now stands on the site. Of the pre-war cinemas which remain standing, the Belgrave is scheduled for demolition, the Plaza is a snooker hall and restaurant, the State is unused, and the Gaumont Palace is currently empty having previously been a nightclub and is undergoing a stalled conversion to a religious centre. In Devonport the Forum, the only cinema there to survive the war, is now a bingo hall.

The Royal Cinema has been threatened with closure on a number of occasions when proposals for its demolition and replacement have been announced. In late 2016, a tower block

End of an Era: Is This the End? Of Course Not, It's Just the Beginning...

Filming the wall of adverts and newspaper articles for BBC Spotlight.

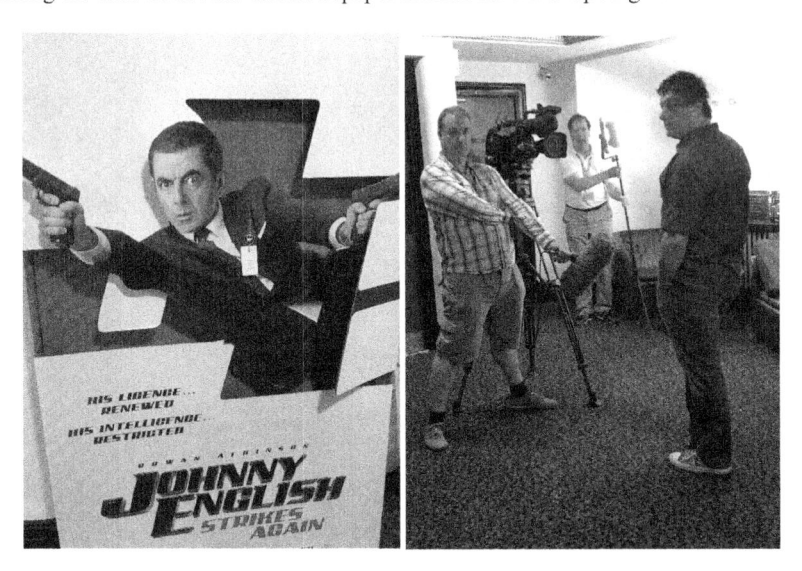

The advertising poster for 'Johnny English Strikes Again'.
Rowan Atkinson is actually wearing one of the 'Reel is Open' hand made tags and my advertising sticker on his lapel, all promotion is good promotion.

The other photo is again the BBC Spotlight crew filming for the News.

End of an Era: Is This the End? Of Course Not, It's Just the Beginning...

Screen 2 on the day of the Anniversary was very quiet in the afternoon.

The BBC Crew arrived to interview staff at the Reel Cinema, including Paul, the long standing Projectionist prior to the introduction of the digital projectors, which no longer required a Projectionist on site.

End of an Era: Is This the End? Of Course Not, It's Just the Beginning...

The Balloons were up to celebrate the Anniversary.

End of an Era: Is This the End? Of Course Not, It's Just the Beginning...

This was the set up for the 80th Cinema Birthday Celebrations and where we were selling copies of the book – 'The Last Picture House: Saving Plymouth's Last Original Cinema Building' (and some others).

On the day, some people came specifically to look around the building and buy copies of the Book and speak with us about our campaign, it was all good publicity and was also covered by a team from the BBC Spotlight programme as well.

End of an Era: Is This the End? Of Course Not, It's Just the Beginning...

Pen & Ink Publishing, Plymouth

with student accommodation and a hotel was proposed, and a year later plans appeared for a 'co-living' scheme, which once the neoliberal gloss is peeled appears to be a way to exploit those failed by a broken housing market, with a very 21st century slum. The Royal is unique in Plymouth- the only pre-war cinema to still operate, and indeed it is unique nationally as the only WR Glen cinema (out of a total of 67) to have operated continuously as a cinema since it was built. Long may the last picture house continue to serve the people of Plymouth."

Karl Parsons (Plymouth)

Bibiography
Atwell D. (1980) **'Cathedrals of the Movies'** –
The Architectural Press, London

Chapman G. (2000) **'Cinema in Devon'**
Devon Books, Tiverton

Eyles A. (1993) **'ABC The First Name in Entertainment'**
BFI Publishing, London

Gould J. (2010) **'Plymouth Vision of a Modern City'** English Heritage, Swindon

Gray R. (1996) **'Cinemas in Britain'**
Lund Humphries Publishers, London

Harwood E. (1994) **'The Listing of Cinemas in English Heritage Conservation Bulletin'**
Vol 22 pp8-9

Paton Watson J. & Abercrombie P. (1943) **'A Plan For Plymouth'**
Underhill, Plymouth

Twyford H.P. (revised by Robinson C.) 2005
'It Came to Our Door'

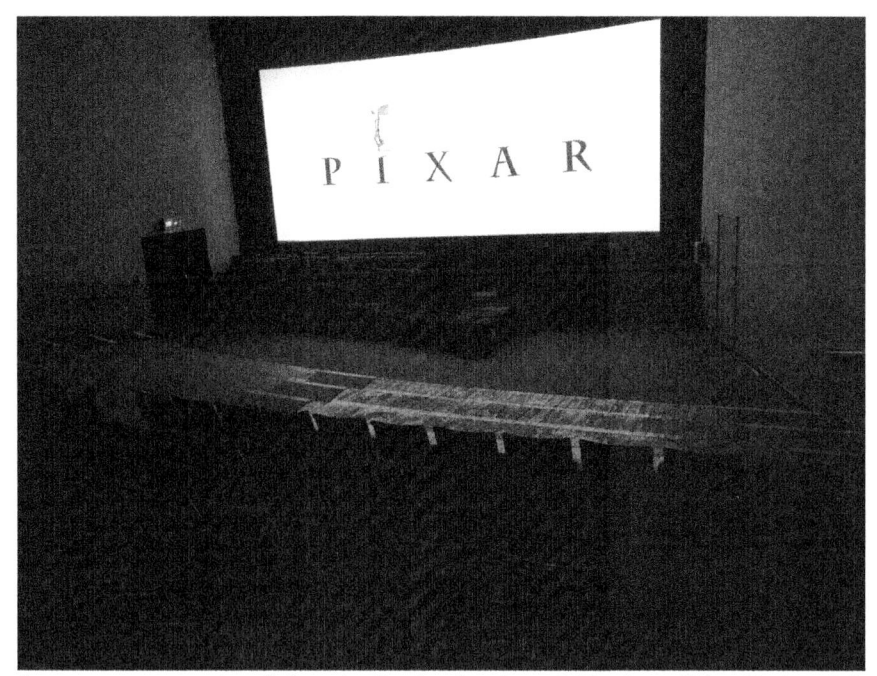

End of an Era: Is This the End? Of Course Not, It's Just the Beginning...

Screen one was again a little quiet for the showing of 'The Incredibles 2', though it was in July and was a beautiful day, people were more than likely out with their families and not wanting to be indoors.

Screen 2, with 2 people waiting for the film to start.

End of an Era: Is This the End? Of Course Not, It's Just the Beginning...

The Cinema plaque that was made for when the cinema celebrated its 50th Anniversary, 30 years ago – back when it was still called the Cannon.

Screen 1 was fairly empty all afternoon, sadly.

End of an Era: Is This the End? Of Course Not, It's Just the Beginning...

The Digital Projector for Screen 1, which saw the releasing of Paul the projectionist from his job, but this was and is the shape of things to come, where digital processes remove manual work and so make people redundant due to not being required anymore, due to the slick and efficient new process.

End of an Era: Is This the End? Of Course Not, It's Just the Beginning...

The Reel Cinemas original 70mm projection equipment, no longer needed, but highly sought after by collectors.
1500 of these were produced by the 'Electro-Acoustics' Division of Philips between 1954 and 1968 and in 1963 it became the only Projector to win an Oscar!

End of an Era: Is This the End? Of Course Not, It's Just the Beginning...

An original 'film splicer' which would be used to repair damaged films and add reels together.

The Battery Room on the roof of the cinema.

Up on the roof of the cinema showing the various rooms that are up there and hopefully will get to be used in the future for various projects. The Civic Centre can be seen in the background; the Civic Centre is already a 'Listed Building', sadly the cinema isn't (yet), despite our efforts already.

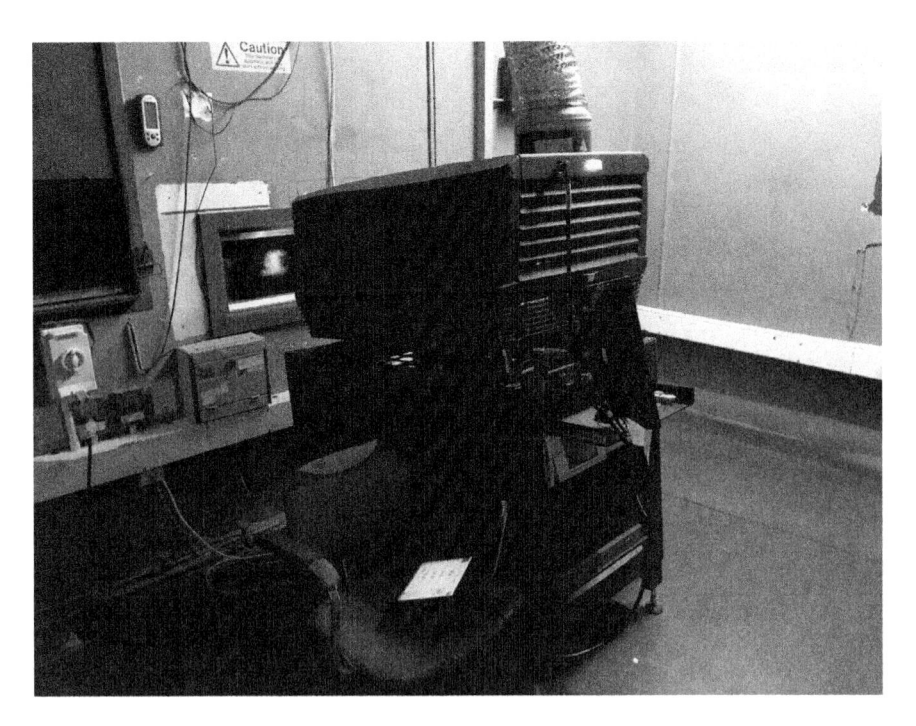

End of an Era: Is This the End? Of Course Not, It's Just the Beginning...

Another view of one of the digital projectors, showing a film in Screen 2.

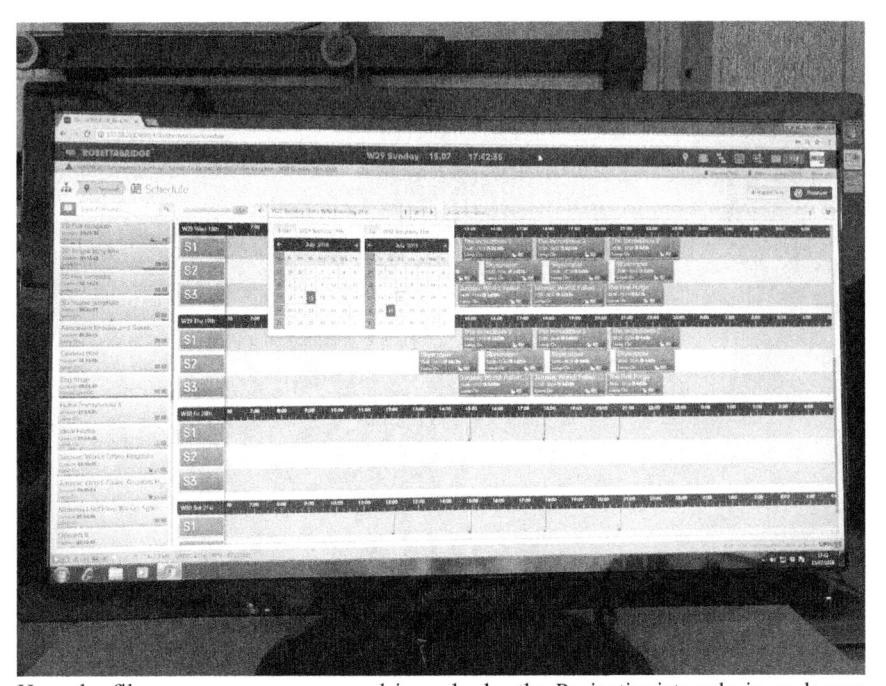

How the films are now programmed in and why the Projectionists role is no longer required.

End of an Era: Is This the End? Of Course Not, It's Just the Beginning...

The 'Old and the New' standing side by side at the projection window for Screen 1.

Looking down into Screen 1 from the Projection room.

End of an Era: Is This the End? Of Course Not, It's Just the Beginning...

One of the views from the roof, on a sunny day in July 2018.

The stairs leading up to the auditorium for Screen 1.

End of an Era: Is This the End? Of Course Not, It's Just the Beginning...

The framework under the seats in Screen 1, which withstood the Luftwaffe and the bombing of Plymouth City Centre.

The severe flooding in the basement and the solid wall structure, looking up to the roof.

All original equipment in a building that has survived so long. The motors on the front are newer, but all the rest is original.

End of an Era: Is This the End? Of Course Not, It's Just the Beginning...

One of the original dressing rooms as the back of the cinema, which has the old antique radiators and classic sinks – these would have been used as changing rooms for theatre performances and also when artists like the Beatles, the Who and Elton John played here.

End of an Era: Is This the End? Of Course Not, It's Just the Beginning...

The view from one of the dressing rooms out onto the street.

An original sink in one of the many dressing rooms.

End of an Era: Is This the End? Of Course Not, It's Just the Beginning...

No words needed…

A little bit of peeling paint, but it is far from damaged, just needs TLC.

End of an Era: Is This the End? Of Course Not, It's Just the Beginning...

The dressing room walls on at the side of the cinema, above the Gala.

Another of the many Dressing Rooms.

End of an Era: Is This the End? Of Course Not, It's Just the Beginning...

The view onto the car park and towards the Crown Court.

The outside and the inside of the building, made of solid brickwork –
so much wasted space that can be utilised in the future for many different things,
ideally for music as the areas are soundproofed.

End of an Era: Is This the End? Of Course Not, It's Just the Beginning...

Items from a time gone by, still lying on the tables in the rooms upstairs, under a covering of dust.

Bizarre antlers and 'balls' combo – any ideas what it is?

End of an Era: Is This the End? Of Course Not, It's Just the Beginning...

Many more strange items in amongst the light bulbs, dust and paper.

All it needs is a lick of paint and a dusting.

End of an Era: Is This the End? Of Course Not, It's Just the Beginning...

More rooms above and at the back of the Reel Cinema.

More peeling paint and dust, but with a dash of paint this will look like new. Another room that will be able to be used for many purposes.

End of an Era: Is This the End? Of Course Not, It's Just the Beginning...

Original artwork on the stairs showing the bands and performers which way that they need to go.

More 80th Anniversary Celebrations.

End of an Era: Is This the End? Of Course Not, It's Just the Beginning...

The Upper Floor during the 80th Celebrations day.

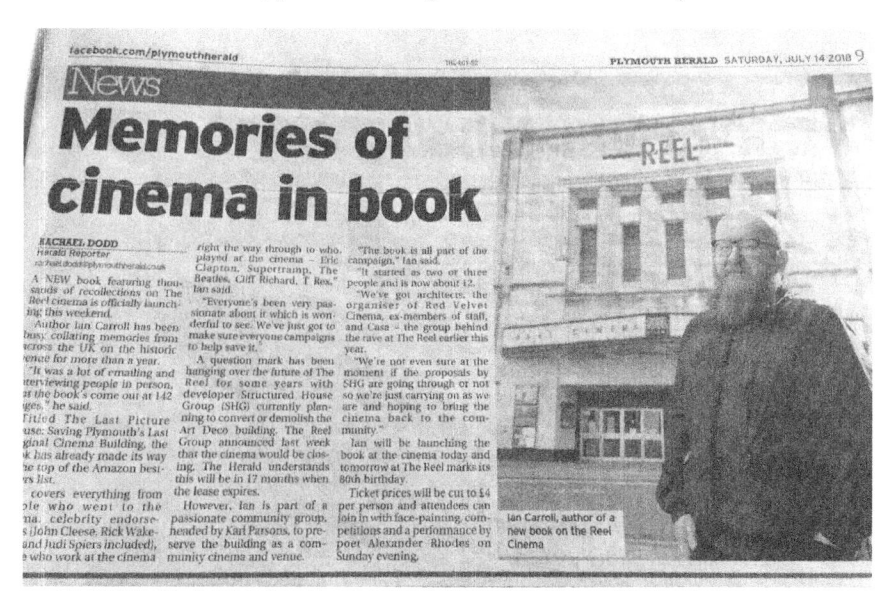

When I was in the Plymouth Herald for the Cinema Book.

"Everyone's been very passionate about it which is wonderful to see. We've just got to make sure everyone campaigns to help save it."

Ian Carroll has written a new book on the Reel Cinema (Image: Penny Cross)

Me, on the Plymouth Live FB page and our message to the Public of Plymouth, who seemed to think that the cinema had already closed.

Spotlight covering the Reel and the Celebrations and the proposed closure - at the time we didn't really know when that would be, it was sooner than we thought.

End of an Era: Is This the End? Of Course Not, It's Just the Beginning...

Paul Gregory - the long time Projectionst talking about the Reel Cinema on Spotlight.

40 years, man and boy…

End of an Era: Is This the End? Of Course Not, It's Just the Beginning...

CHERRIE HARVEY
Box office staff, Reel Cinema
BBC SPOTLIGHT   Email us: spotlight@bbc.co.uk

Cherrie on Spotlight talking about her time at the Reel – which covers a long time. Check the previous book out for the stories that Cherrie tells about her long career at the venue.

Dani (the Manager) and Paul on Spotlight, with Caroline from the Royal Cinema Trust in the background.

End of an Era: Is This the End? Of Course Not, It's Just the Beginning...

A signing session for 'The Last Picture House' Book at the Hive Mind, Plymouth's most amazing independent comic book shop, with a massive collection of Funko Pop figures for sale too.

Posters advertising both the 80th Anniversary at the cinema and the fund raising concert that we put on at the Underground, Mutley Plain, Plymouth with Young Nuns, Roz Birch and the Waterboarders, which was a big success and we spread the word about the campaign and raise more funds.

End of an Era: Is This the End? Of Course Not, It's Just the Beginning...

The poster on the door for the concert, if you pass the Underground to this day it's still there, though a little faded and worse for wear, just like the paintwork in the rooms above the cinema. We collecting donations for entry and for a Thursday evening it was quite busy.

End of an Era: Is This the End? Of Course Not, It's Just the Beginning...

The Young Nuns @ the Underground for the Fundraiser

End of an Era: Is This the End? Of Course Not, It's Just the Beginning...

Roz Birch @ the Underground for the Fundraiser

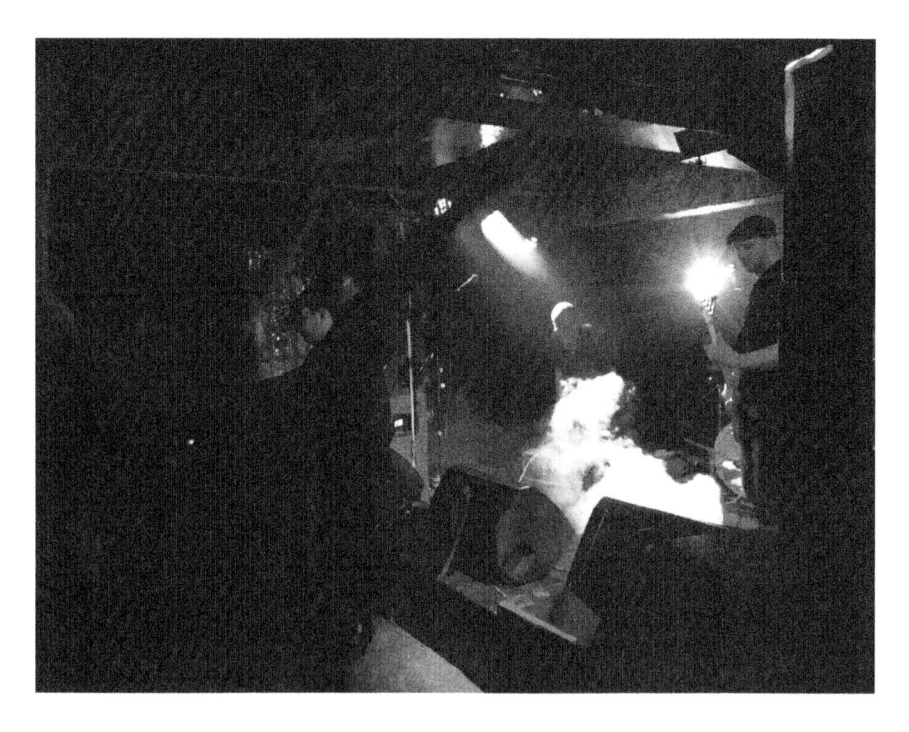

End of an Era: Is This the End? Of Course Not, It's Just the Beginning...

The Waterboarders @ the Underground for the Fundraiser

End of an Era: Is This the End? Of Course Not, It's Just the Beginning...

The Gala Bingo Hall, now emptied, will make an amazing sized concert venue, which could hold approximately 1,200 people easily.

End of an Era: Is This the End? Of Course Not, It's Just the Beginning...

Looking out from the stage at the Gala Bingo, what bands could play here in the future? Who knows, but we live in hope...

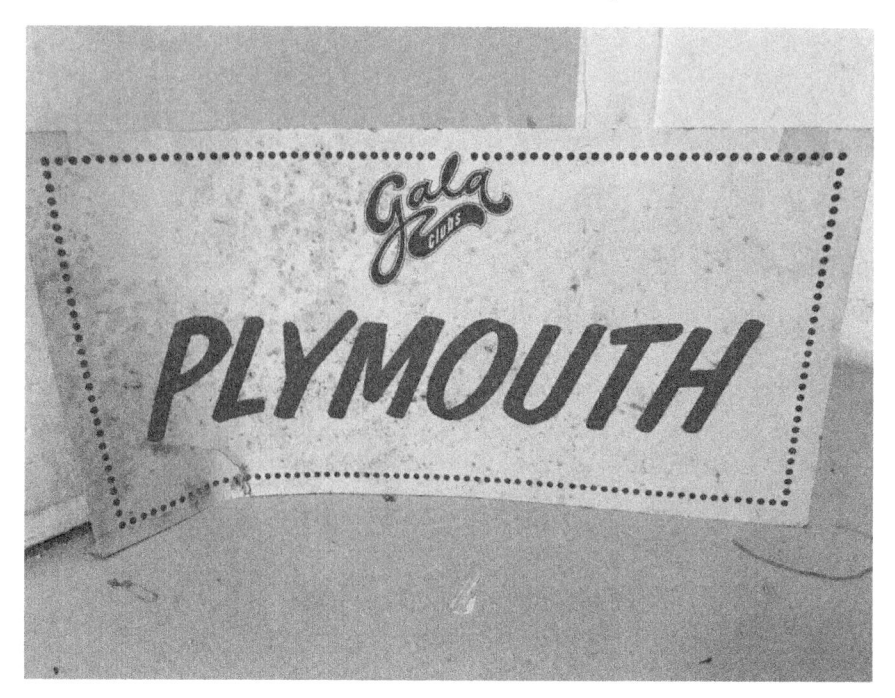

End of an Era: Is This the End? Of Course Not, It's Just the Beginning...

Above the ceiling in the Gala Bingo Hall

Even the 'chalk based' graffiti is on our side, join the campaign.

End of an Era: Is This the End? Of Course Not, It's Just the Beginning...

All the original ceilings are above the false ceilings in the Gala, this artwork is the underside of the original cinema circle, which ended up as screen 1 in the cinema.

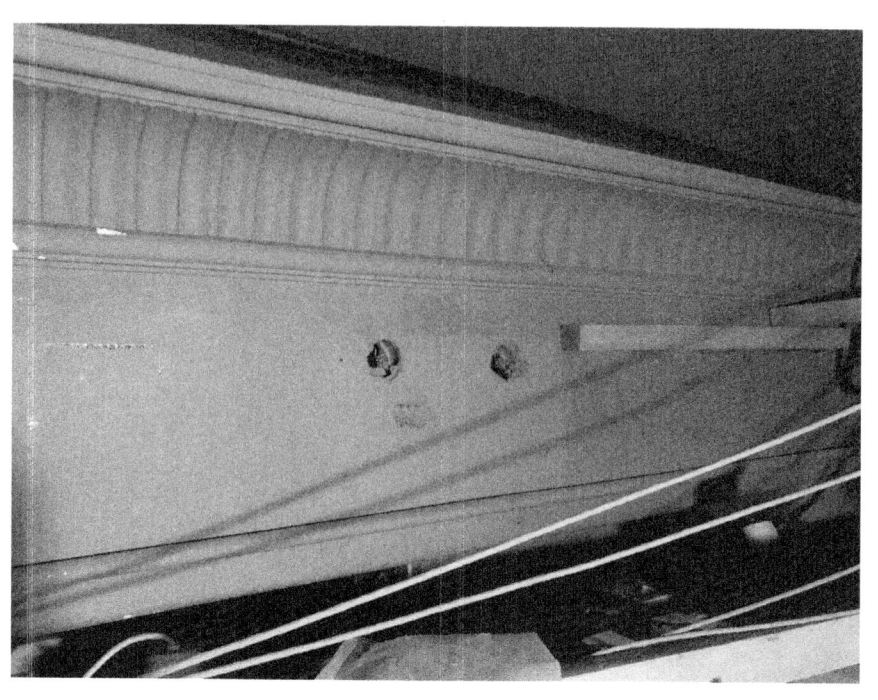

End of an Era: Is This the End? Of Course Not, It's Just the Beginning...

More of the original ceiling above the Gala Bingo Hall.

End of an Era: Is This the End? Of Course Not, It's Just the Beginning...

More of the original plasterwork above the false ceiling.

Up on the roof, which would make a great 'rooftop' bar.

End of an Era: Is This the End? Of Course Not, It's Just the Beginning...

Above the screens in the Cinemas roof, an incredibly solid construction.

End of an Era: Is This the End? Of Course Not, It's Just the Beginning...

Girders, girders and more girders above the screens.

End of an Era: Is This the End? Of Course Not, It's Just the Beginning...

Amazingly solid construction that was one of the few buildings in the area to survive the bombings intact.

End of an Era: Is This the End? Of Course Not, It's Just the Beginning...

It's a great view from the roof, looking towards the Theatre Royal.

It's quite a height from the roof-space at the top of the cinema building.

End of an Era: Is This the End? Of Course Not, It's Just the Beginning...

Original Art Deco glasswork in the door to the Gentlemen's toilet.

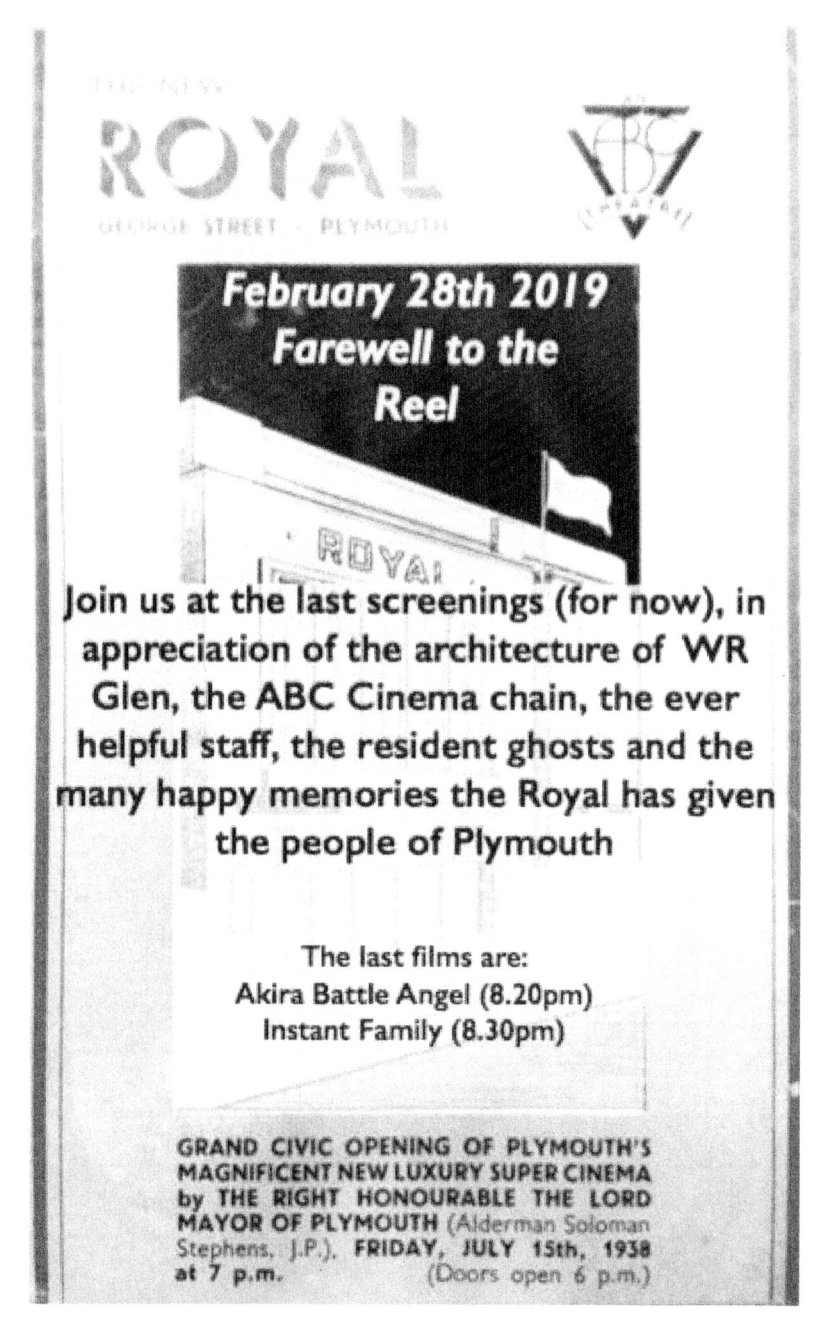

End of an Era: Is This the End? Of Course Not, It's Just the Beginning...

THE NEW ROYAL
GEORGE STREET · PLYMOUTH

February 28th 2019
Farewell to the Reel

Join us at the last screenings (for now), in appreciation of the architecture of WR Glen, the ABC Cinema chain, the ever helpful staff, the resident ghosts and the many happy memories the Royal has given the people of Plymouth

The last films are:
Akira Battle Angel (8.20pm)
Instant Family (8.30pm)

GRAND CIVIC OPENING OF PLYMOUTH'S MAGNIFICENT NEW LUXURY SUPER CINEMA by THE RIGHT HONOURABLE THE LORD MAYOR OF PLYMOUTH (Alderman Soloman Stephens, J.P.). FRIDAY, JULY 15th, 1938 at 7 p.m.   (Doors open 6 p.m.)

Adverts produced by the Royal Cinema Trust for the Reel's final evening.

End of an Era: Is This the End? Of Course Not, It's Just the Beginning...

Karl being interviewed for a film about Plymouth's buildings that are being demolished/taken away.

End of an Era: Is This the End? Of Course Not, It's Just the Beginning...

Being interviewed for the Documentary as well.

End of an Era: Is This the End? Of Course Not, It's Just the Beginning...

The Warning Sign outside of Screen 1

End of an Era: Is This the End? Of Course Not, It's Just the Beginning...

Hundreds of seats and seat backs, stored in the back of the Reel.

Outside the Main entrance, showing films that would never be shown.

A naked hand-drier in one of the 'out of use' toilets.

More empty rooms at the back of the Cinema.

End of an Era: Is This the End? Of Course Not, It's Just the Beginning...

What ghosts have wandered these eerie corridors?

It's terrifying in the partial lighted corridors.

End of an Era: Is This the End? Of Course Not, It's Just the Beginning...

The Reel Building looking dwarfed by the new tower blocks.

End of an Era: Is This the End? Of Course Not, It's Just the Beginning...

The men's toilets on the final evening of the Reel Cinema.

End of an Era: Is This the End? Of Course Not, It's Just the Beginning...

The films are all over and it's nearly time to leave the Reel Cinema.

Screen 1 looking very sad on the last day, after everyone had left.

End of an Era: Is This the End? Of Course Not, It's Just the Beginning...

Alone in Screen 1 – sad times…

End of an Era: Is This the End? Of Course Not, It's Just the Beginning...

What's most sad than the end of the foyer? Gone are all the sweets, the popcorn, the hot dogs and the drinks are ready to be packed away…

End of an Era: Is This the End? Of Course Not, It's Just the Beginning...

> THANK YOU
> FOR ALL YOUR SUPPORT
> OVER THE YEARS
>
> WE CLOSE OUR DOORS
> FOR THE LAST TIME
> AFTER THE LAST SHOW
> THE 28TH FEBRUARY

That's made it all seem so final.

End of an Era: Is This the End? Of Course Not, It's Just the Beginning...

The following morning I went to the BBC in Seymour Road, Plymouth and featured on the Breakfast Show, talking about the cinema and what did we see in the future for the building. I think I did well enough and covered all points that were relevant with our campaign.

Good old BBC, to feature our campaign on
BBC Radio Devon.

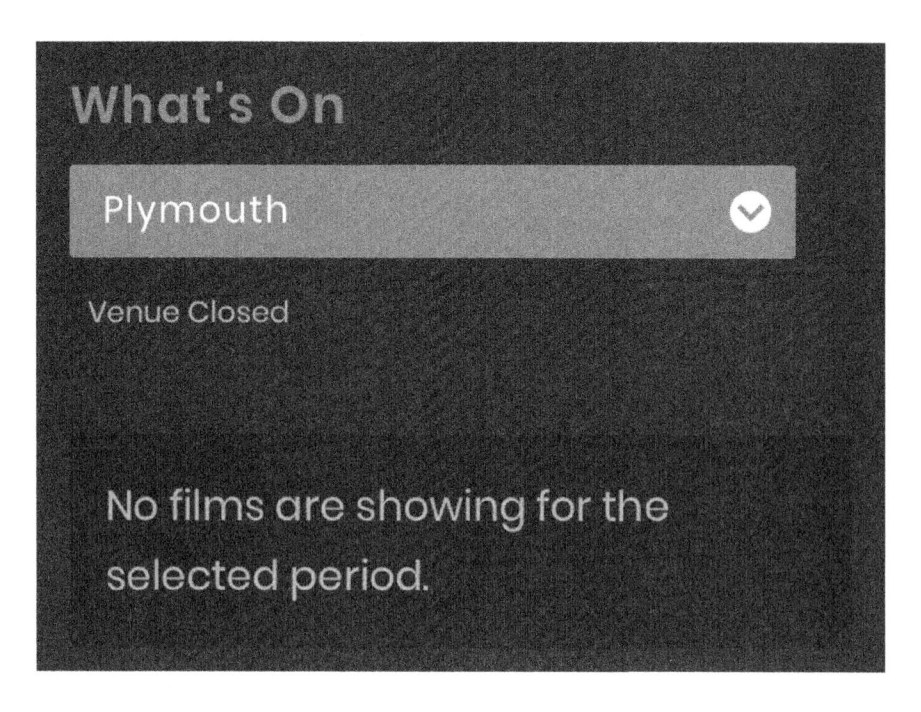

End of an Era: Is This the End? Of Course Not, It's Just the Beginning...

> **What's On**
>
> Plymouth
>
> Venue Closed
>
> No films are showing for the selected period.

The day after the Reel had closed and their website puts the final nail in the coffin.

A Visit to the Reel two days after the closure.

End of an Era: Is This the End? Of Course Not, It's Just the Beginning…

The Reel Cinema, in amongst the tower blocks.

All boarded up – no more films, no more Saturday Morning shows…

End of an Era: Is This the End? Of Course Not, It's Just the Beginning...

Closed for the moment.

The reusable stock is boxed, bagged and ready to go.

End of an Era: Is This the End? Of Course Not, It's Just the Beginning...

All being sent away to other Reel Cinemas, that are still open.

From behind the till.

End of an Era: Is This the End? Of Course Not, It's Just the Beginning...

The whole building boarded up, only Lorenzos is still trading.

The warning to the public about breaking in.

End of an Era: Is This the End? Of Course Not, It's Just the Beginning...

Screen 3 looking sad, lonely and empty.

Screen 3, still ready to go, last film that I saw there was 'John Wick 2'

End of an Era: Is This the End? Of Course Not, It's Just the Beginning...

Such a waste of a beautiful cinema space.

The empty upstairs area, no punters, no staff, nothing.

End of an Era: Is This the End? Of Course Not, It's Just the Beginning...

Screen 1 where I saw 'Alita Battle Angel' – the last film I watched here.

End of an Era: Is This the End? Of Course Not, It's Just the Beginning...

Screen 1, looking ready to go. The seats are clean and the walls are spotless – such a waste of a wonderful venue.

End of an Era: Is This the End? Of Course Not, It's Just the Beginning...

The stairs just need a 'lick of paint' and then it'll be ready for the next performance, whenever that may be.

Screen 2, seats like new.

End of an Era: Is This the End? Of Course Not, It's Just the Beginning...

Sad times…

End of an Era: Is This the End? Of Course Not, It's Just the Beginning...

Karl *'Father of Two'* Parsons leading the campaign to save the building.

Ian Carroll outside the Reel for the Book for the first Book on the Cinema, but the campaign isn't over, as long as the building is still standing.

End of an Era: Is This the End? Of Course Not, It's Just the Beginning...

Danielle McCourt the final Manager
at the Reel Cinema

The Reel Building, hiding in amongst the surrounding larger
buildings, it's still there – JOIN the Campaign.

End of an Era: Is This the End? Of Course Not, It's Just the Beginning...

Epitaph

So, the cinema has closed, the plans for the building still seem to be very *'up in the air'* with no immediate plans for the demolition of the building, we can still save it and it can still be a community venue for the people.

So join up at the Facebook page –

Save the Royal (ABC/Reel Cinema Building)

We already have a thriving page with 673 people having already joined the group and more joining each week. Read the page for stories about other cinema buildings that have been saved and plenty of stories about what we are up to with the campaign.

It ain't over till the fat lady sings…

Ian Carroll

The Last Picture House: Saving Plymouth's Last Original Cinema Building shared a link.
22 March 2017

Printed in Great Britain
by Amazon